The Official Sassafras SCIDAT Logbook

Earth Science Edition

Prepared by: _____

The Official Sassafras SCIDAT Logbook: Earth Science Edition

First Edition 2016
Copyright @ Elemental Science, Inc.
Email: info@elementalscience.com

ISBN #978-1-935614-44-9

Printed In USA For World Wide Distribution

For more copies write to:
Elemental Science
610 N. Main St., #207
Blacksburg, VA 24060
info@elementalscience.com

Copyright Policy

All contents copyright © 2016 by Elemental Science. All rights reserved.

No part of this document or the related files may be reproduced or transmitted in any form, by any means (electronic, photocopying, recording, or otherwise) without the prior written permission of the author. The author does give permission to the original purchaser to photocopy all supplemental material for use within their immediate family only.

Limit of Liability and Disclaimer of Warranty: The publisher has used its best efforts in preparing this book, and the information provided herein is provided "as is." Elemental Science makes no representation or warranties with respect to the accuracy or completeness of the contents of this book and specifically disclaims any implied warranties of merchantability or fitness for any particular purpose and shall in no event be liable for any loss of profit or any other commercial damage, including but not limited to special, incidental, consequential, or other damages.

Trademarks: This book identifies product names and services known to be trademarks, registered trademarks, or service marks of their respective holders. They are used throughout this book in an editorial fashion only. In addition, terms suspected of being trademarks, registered trademarks, or service marks have been appropriately capitalized, although Elemental Science cannot attest to the accuracy of this information. Use of a term in this book should not be regarded as affecting the validity of any trademark, registered trademark, or service mark. Elemental Science is not associated with any product or vendor mentioned in this book.

The Official Sassafras SCIDAT Logbook: Earth Science Edition

Table of Contents

Weather Observations..5

Oklahoman Prairie..7

Congolese Rainforest..17

Patagonian Mountains...29

Mongolian Desert..39

Pakistani Mountains..49

Alaskan Boreal Forest...61

Pacific Ocean...71

Swiss Deciduous Forest..81

Bonus Data..91

Earth Science Glossary...93

Earth Science Notes
Weather Observations

3/26/2020 Warm, sunny, scattered clouds, cool breeze

Narration / Copywork / Dictation

EARTH SCIENCE NOTES
WEATHER OBSERVATIONS

SCIDAT Logbook
Oklahoman Prairie Climate Sheet

Area Map

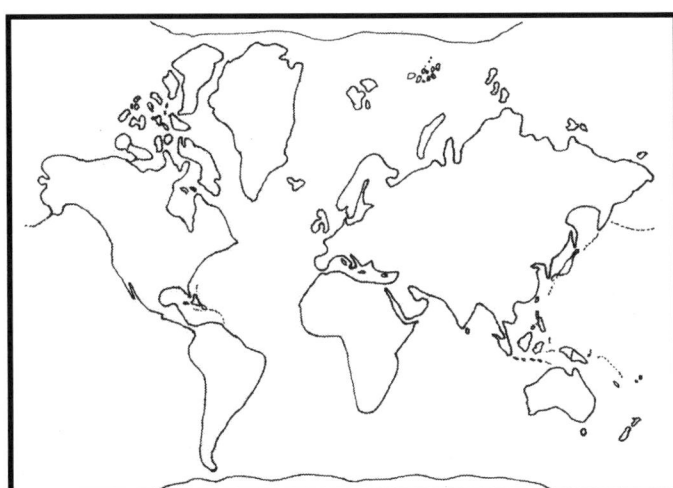

Climate Information

Interesting Facts

Other Types

Weather Records for the Weeks of _____

	Monday	Tuesday	Wednesday	Thursday	Friday
High					
Low					
Rainfall					
Conditions					

	Monday	Tuesday	Wednesday	Thursday	Friday
High					
Low					
Rainfall					
Conditions					

SCIDAT Logbook
Earth Science Record Sheet

Wind

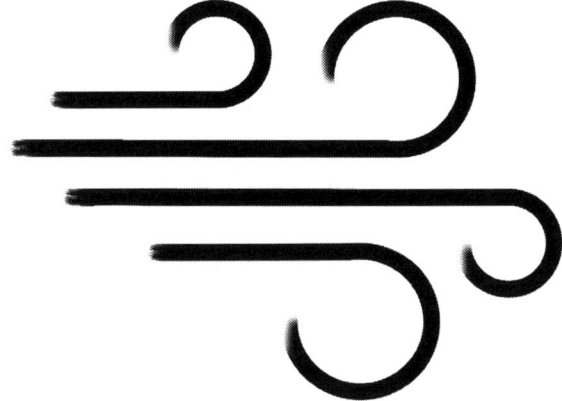

Information Learned

SCIDAT Logbook
Earth Science Record Sheet

Global Wind Patterns

Information Learned

The Official Sassafras SCIDAT Logbook ~ Earth Science Edition

SCIDAT Logbook
Earth Science Record Sheet

Downbursts

Information Learned

SCIDAT Logbook
Earth Science Record Sheet

Tornadoes

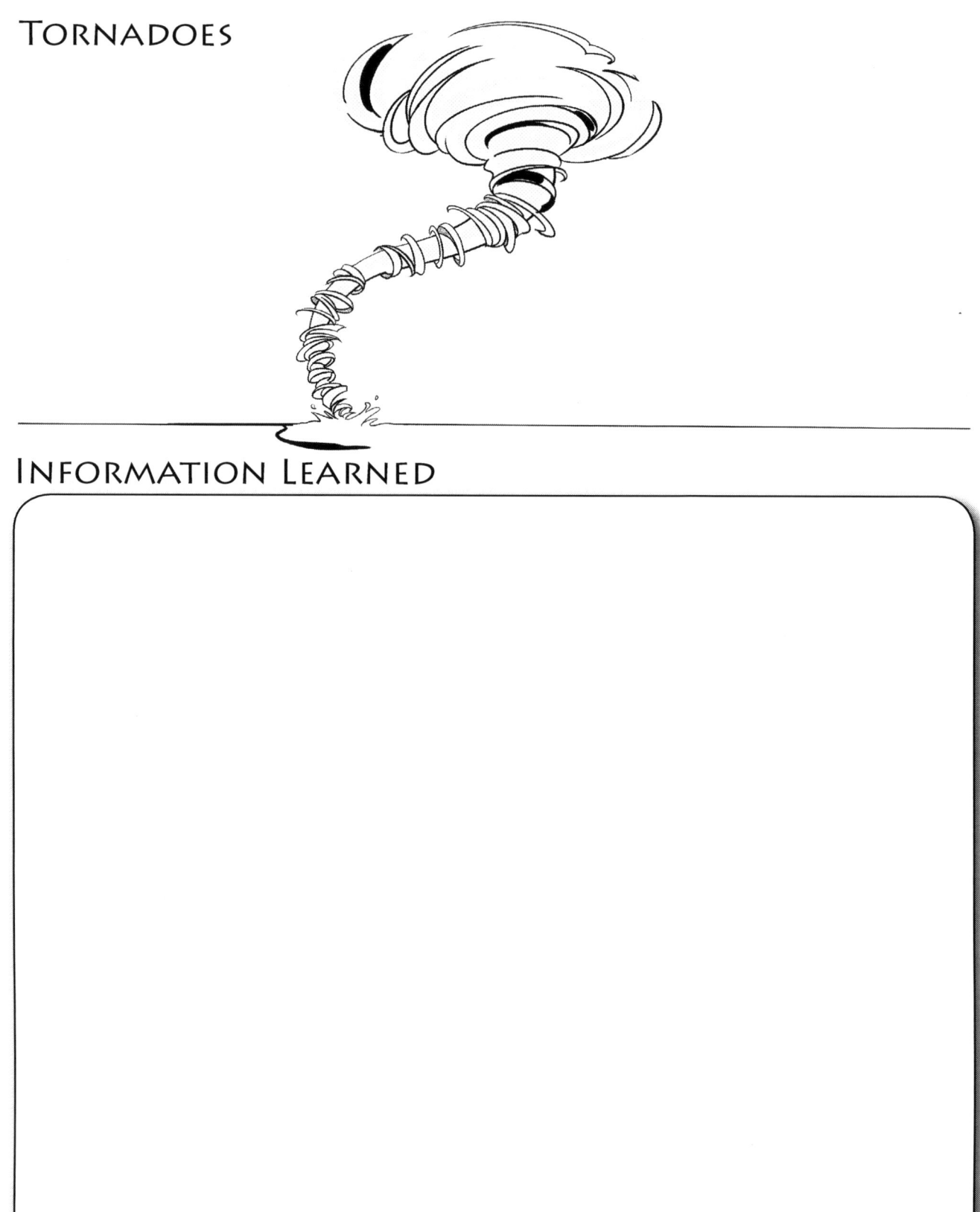

Information Learned

Earth Science Notes
Oklahoman Prairie

Earth Science Notes
Oklahoman Prairie

SCIDAT Logbook
Project Record Sheet

Glue Picture of
Project Here

Information Learned

SCIDAT Logbook
Project Record Sheet

Glue Picture of Project Here

Information Learned

SCIDAT Logbook
Congolese Rainforest Climate Sheet

Area Map

Climate Information

Interesting Facts

Other Types

Weather Records for the Weeks of _____

	Monday	Tuesday	Wednesday	Thursday	Friday
High					
Low					
Rainfall					
Conditions					

	Monday	Tuesday	Wednesday	Thursday	Friday
High					
Low					
Rainfall					
Conditions					

SCIDAT Logbook
Earth Science Record Sheet

Rain

Information Learned

SCIDAT Logbook
Earth Science Record Sheet

Monsoons

Information Learned

SCIDAT Logbook
Earth Science Record Sheet

Thunderstorm

Information Learned

SCIDAT Logbook
Earth Science Record Sheet

Flood

Information Learned

Earth Science Notes
Congolese Rainforest

Earth Science Notes
Congolese Rainforest

Rain Gauge Project

Glue Picture of Project Here

Week 1

Day	1	2	3	4	5	6	7
Predicted rainfall*							
Actual rainfall*							

*Don't forget to include units like inches (in) or centimeters (cm) with your measurements.

Week 2

Day	1	2	3	4	5	6	7
Predicted rainfall*							
Actual rainfall*							

*Don't forget to include units like inches (in) or centimeters (cm) with your measurements.

Rain Gauge Project

Glue Picture of Project Here

Week 3

Day	1	2	3	4	5	6	7
Predicted rainfall*							
Actual rainfall*							

*Don't forget to include units like inches (in) or centimeters (cm) with your measurements.

Week 4

Day	1	2	3	4	5	6	7
Predicted rainfall*							
Actual rainfall*							

*Don't forget to include units like inches (in) or centimeters (cm) with your measurements.

SCIDAT Logbook
Project Record Sheet

Glue Picture of
Project Here

Information Learned

SCIDAT Logbook
Project Record Sheet

Glue Picture of Project Here

Information Learned

SCIDAT Logbook
Patagonian Mountains Climate Sheet

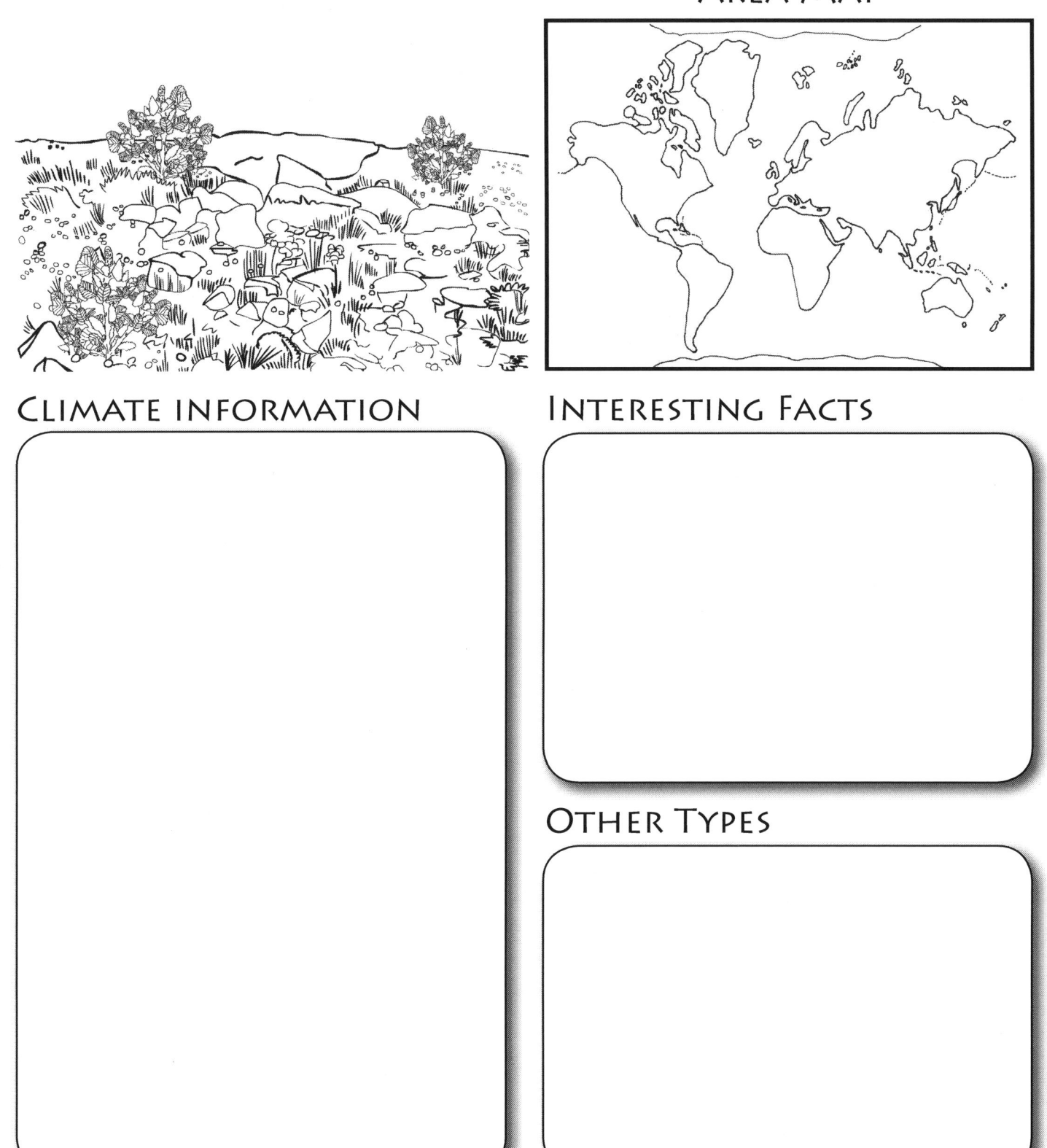

Area Map

Climate Information

Interesting Facts

Other Types

Weather Records for the Weeks of _____

	Monday	Tuesday	Wednesday	Thursday	Friday
High					
Low					
Rainfall					
Conditions					

	Monday	Tuesday	Wednesday	Thursday	Friday
High					
Low					
Rainfall					
Conditions					

SCIDAT Logbook
Earth Science Record Sheet

Snow

Information Learned

SCIDAT Logbook
Earth Science Record Sheet

Ice Storm

Information Learned

SCIDAT Logbook
Earth Science Record Sheet

Frost Quake

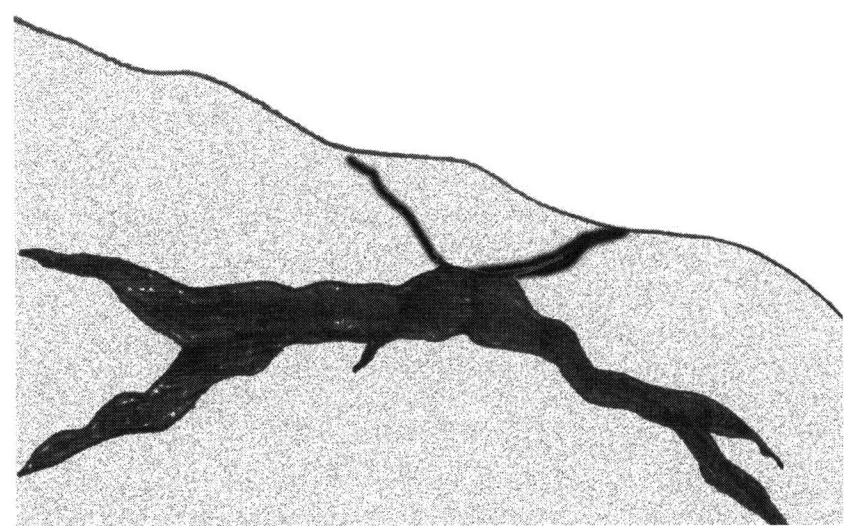

Information Learned

SCIDAT Logbook
Earth Science Record Sheet

Seasons

Information Learned

Earth Science Notes
Patagonian Mountains

Earth Science Notes
Patagonian Mountains

SCIDAT Logbook
Project Record Sheet

Glue Picture of
Project Here

Information Learned

SCIDAT Logbook
Project Record Sheet

Glue Picture of
Project Here

Information Learned

SCIDAT Logbook
Mongolian Desert Climate Sheet

Area Map

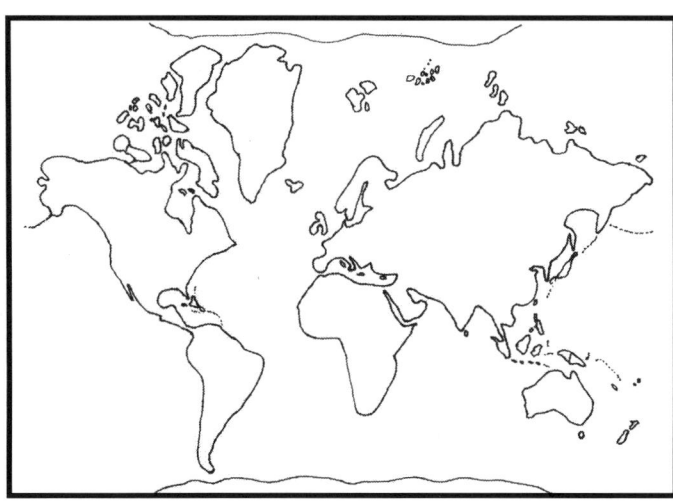

Climate Information

Interesting Facts

Other Types

Weather Records for the Weeks of _____

	Monday	Tuesday	Wednesday	Thursday	Friday
High					
Low					
Rainfall					
Conditions					

	Monday	Tuesday	Wednesday	Thursday	Friday
High					
Low					
Rainfall					
Conditions					

SCIDAT Logbook
Earth Science Record Sheet

Day / Night

Information Learned

SCIDAT Logbook
Earth Science Record Sheet

Sandstorm

Information Learned

SCIDAT Logbook
Earth Science Record Sheet

Drought

Information Learned

SCIDAT Logbook
Earth Science Record Sheet

Oasis

Information Learned

Earth Science Notes
Mongolian Desert

Earth Science Notes
Mongolian Desert

SCIDAT Logbook
Project Record Sheet

Glue Picture of
Project Here

Information Learned

SCIDAT Logbook
Project Record Sheet

Glue Picture of
Project Here

Information Learned

SCIDAT Logbook
Pakistani Mountains Climate Sheet

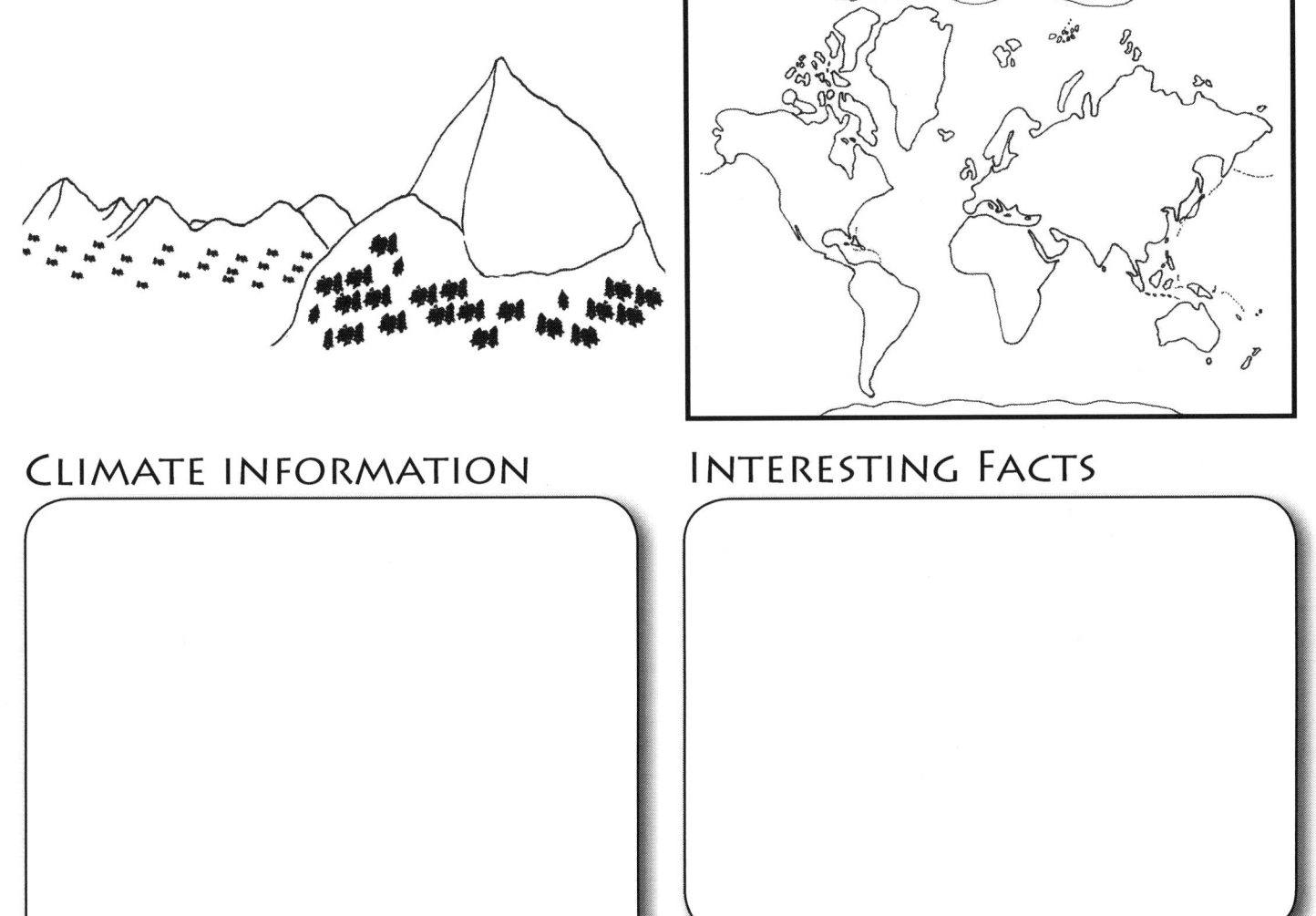

Area Map

Climate Information

Interesting Facts

Other Types

Weather Records for the Weeks of _____

	Monday	Tuesday	Wednesday	Thursday	Friday
High					
Low					
Rainfall					
Conditions					

	Monday	Tuesday	Wednesday	Thursday	Friday
High					
Low					
Rainfall					
Conditions					

SCIDAT Logbook
Earth Science Record Sheet

Atmosphere

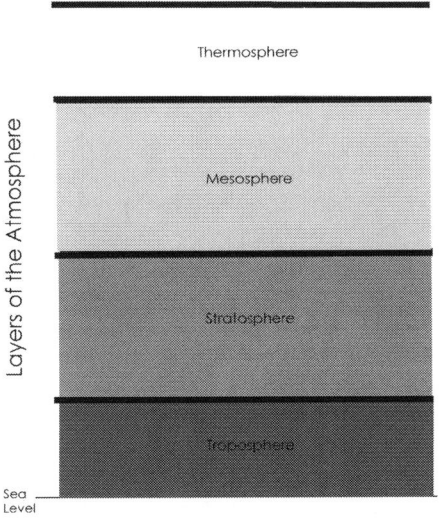

Information Learned

SCIDAT Logbook
Earth Science Record Sheet

Clouds

Information Learned

SCIDAT Logbook
Earth Science Record Sheet

Cirrus Clouds

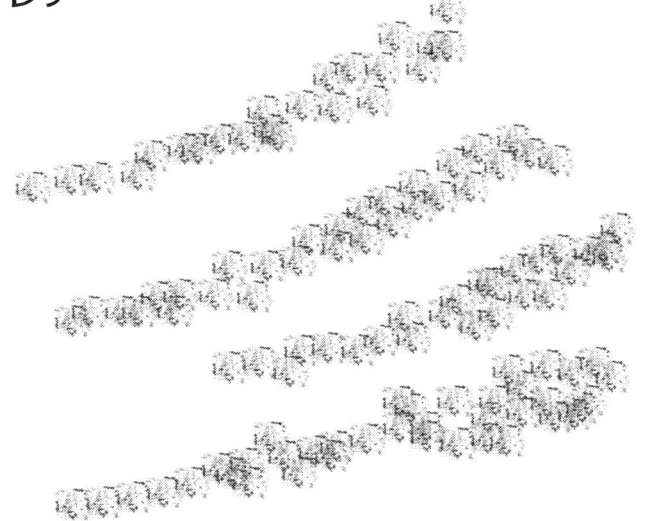

Information Learned

SCIDAT Logbook
Earth Science Record Sheet

Alto Clouds

Information Learned

SCIDAT Logbook
Earth Science Record Sheet

Stratus Clouds

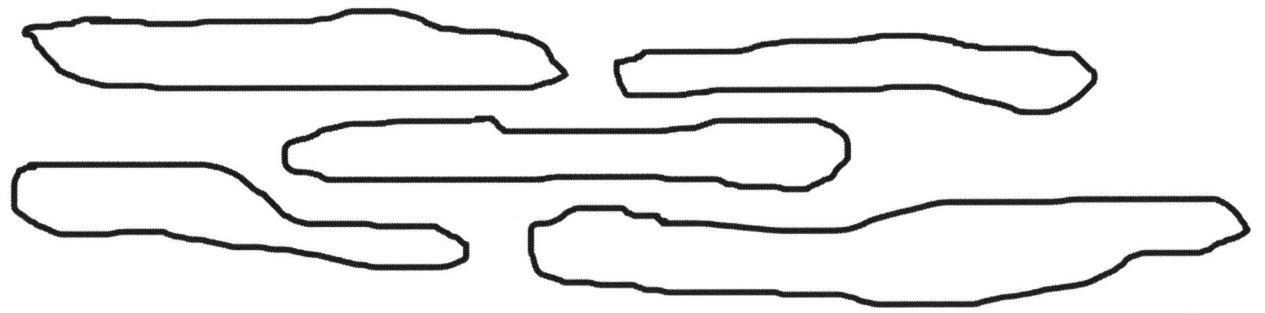

Information Learned

SCIDAT Logbook
Earth Science Record Sheet

Cumulus Clouds

Information Learned

Earth Science Notes
Pakistani Mountains

Earth Science Notes
Pakistani Mountains

SCIDAT Logbook
Project Record Sheet

Glue Picture of
Project Here

Information Learned

SCIDAT Logbook
Project Record Sheet

Glue Picture of
Project Here

Information Learned

SCIDAT Logbook
Alaskan Boreal Forest Climate Sheet

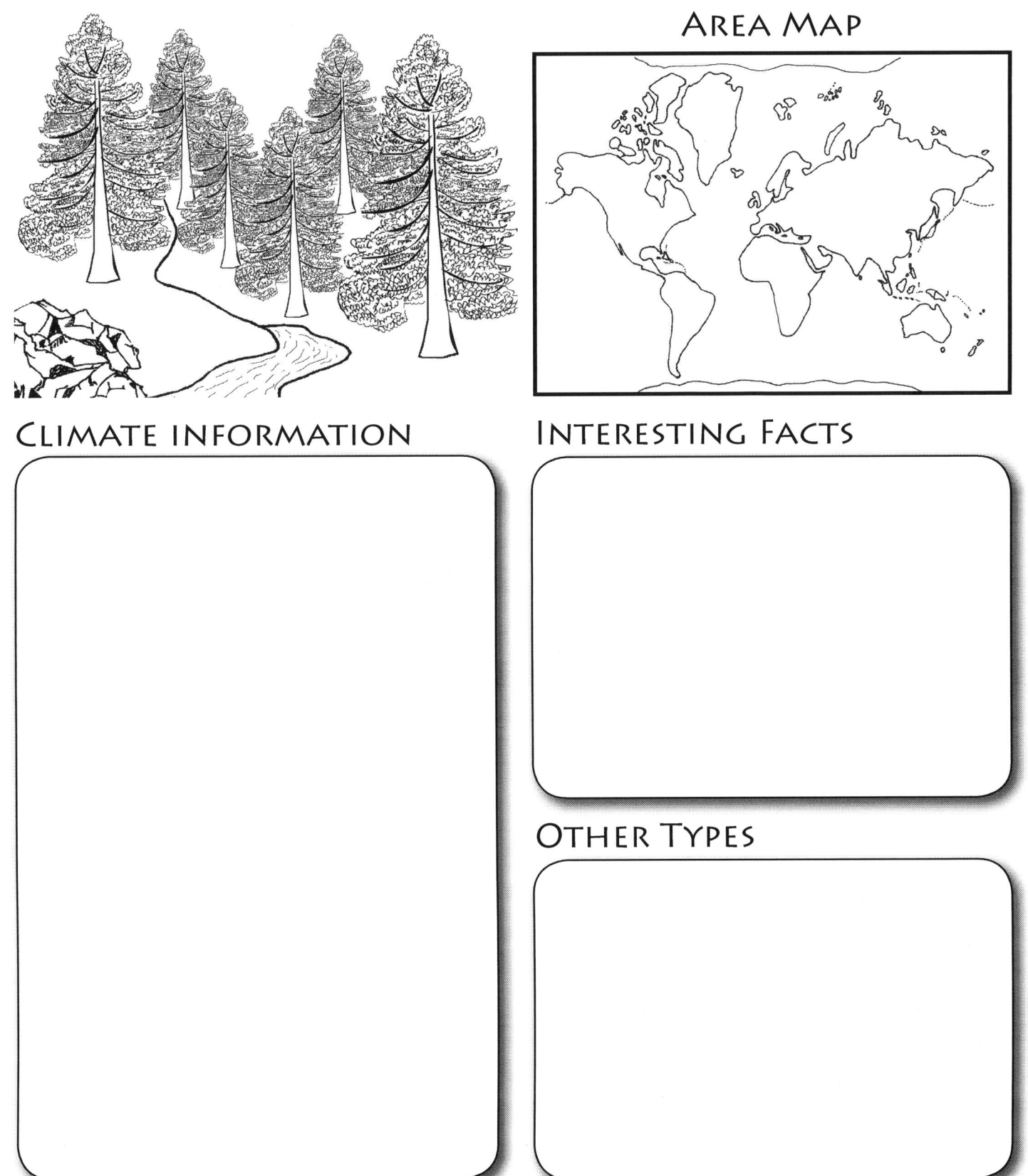

Area Map

Climate Information

Interesting Facts

Other Types

Weather Records for the Weeks of _____

	Monday	Tuesday	Wednesday	Thursday	Friday
High					
Low					
Rainfall					
Conditions					

	Monday	Tuesday	Wednesday	Thursday	Friday
High					
Low					
Rainfall					
Conditions					

SCIDAT Logbook
Earth Science Record Sheet

Water Cycle

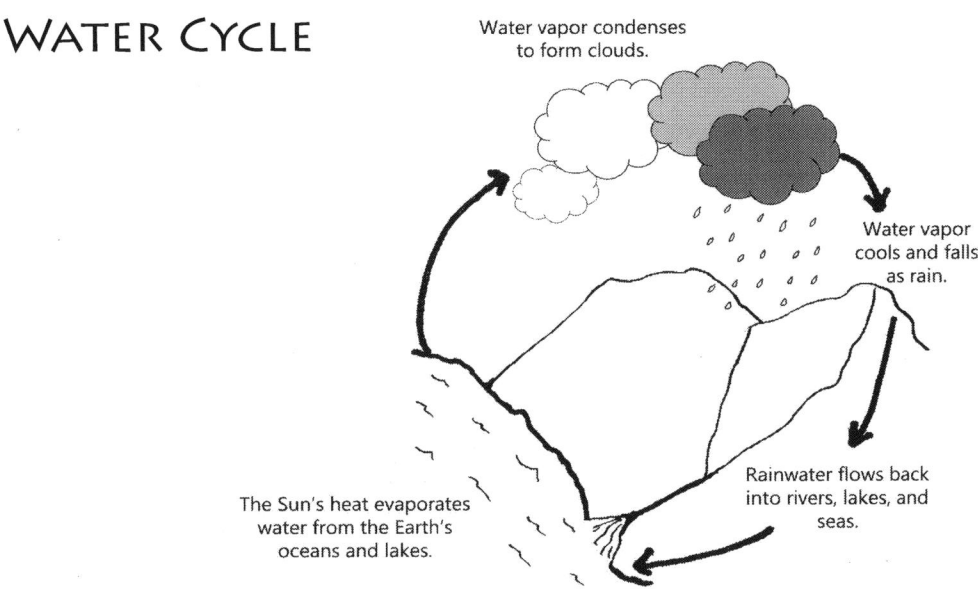

Information Learned

SCIDAT Logbook
Earth Science Record Sheet

Fog

Information Learned

SCIDAT Logbook
Earth Science Record Sheet

Nitrogen cycle

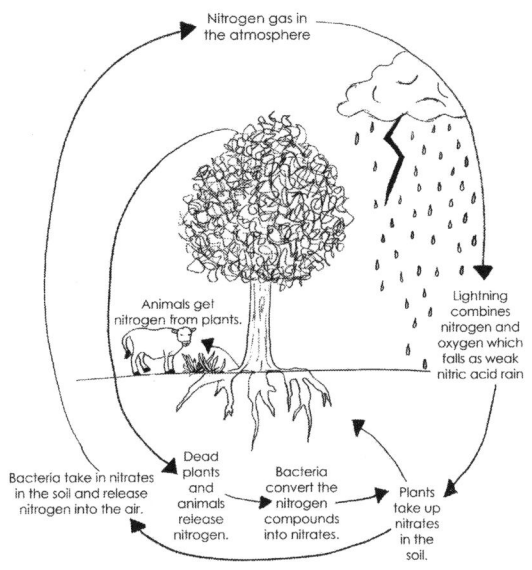

Information Learned

SCIDAT Logbook
Earth Science Record Sheet

Phosphorus Cycle

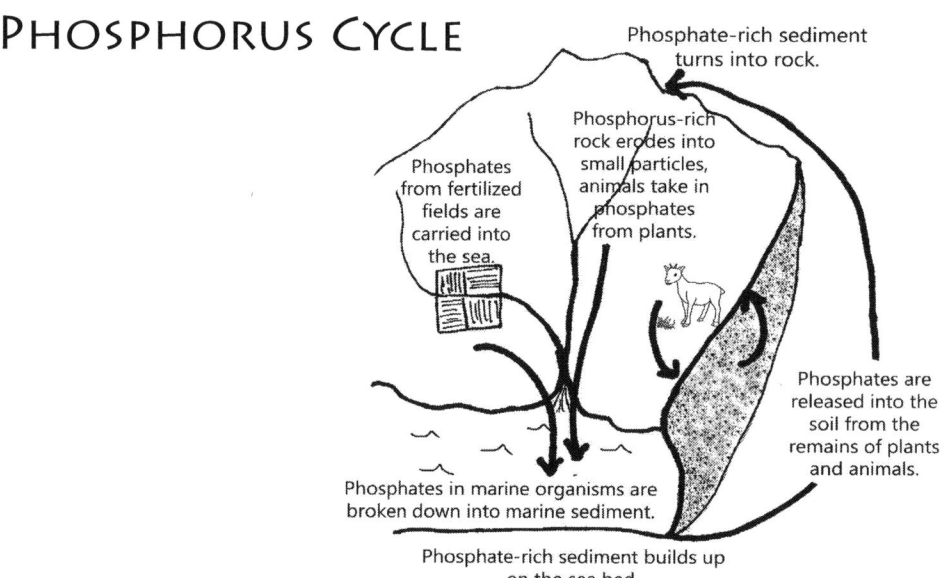

Information Learned

Earth Science Notes
Alaskan Boreal Forest

Earth Science Notes
Alaskan Boreal Forest

SCIDAT Logbook
Project Record Sheet

Glue Picture of
Project Here

Information Learned

SCIDAT Logbook
Project Record Sheet

Glue Picture of
Project Here

Information Learned

SCIDAT Logbook
Pacific Ocean Climate Sheet

Area Map

Interesting Facts

Weather Records for the Weeks of _____

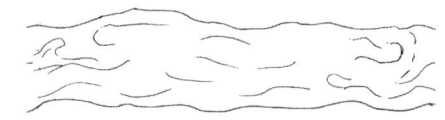

	Monday	Tuesday	Wednesday	Thursday	Friday
High					
Low					
Rainfall					
Conditions					

	Monday	Tuesday	Wednesday	Thursday	Friday
High					
Low					
Rainfall					
Conditions					

SCIDAT Logbook
Earth Science Record Sheet

Coral Reef

Information Learned

SCIDAT Logbook
Earth Science Record Sheet

Currents

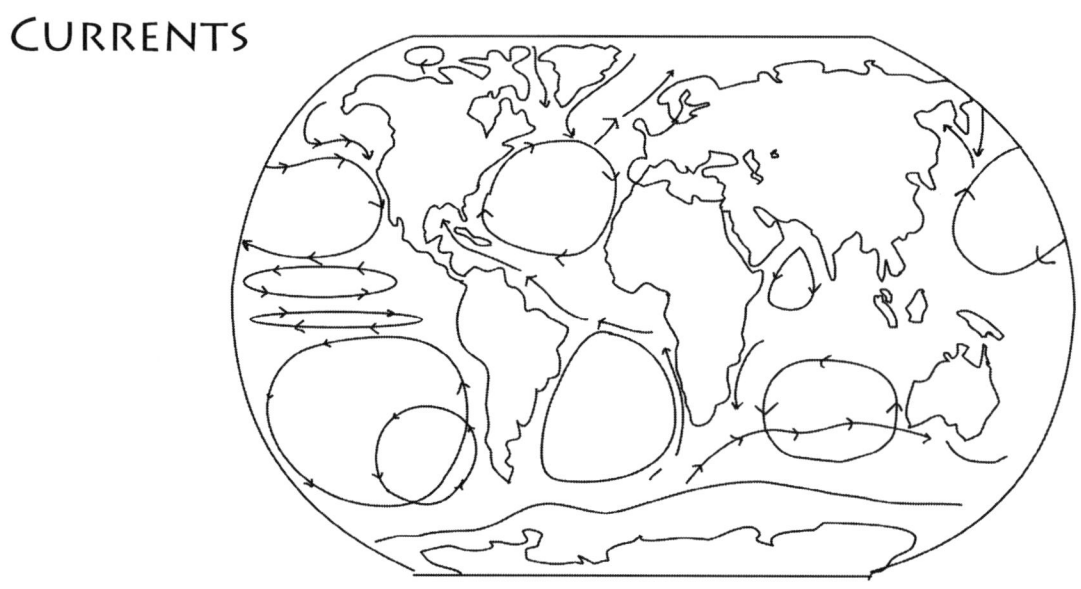

Information Learned

The Official Sassafras SCIDAT Logbook ~ Earth Science Edition

SCIDAT Logbook
Earth Science Record Sheet

Oceans

Information Learned

75

SCIDAT Logbook
Earth Science Record Sheet

Hurricanes

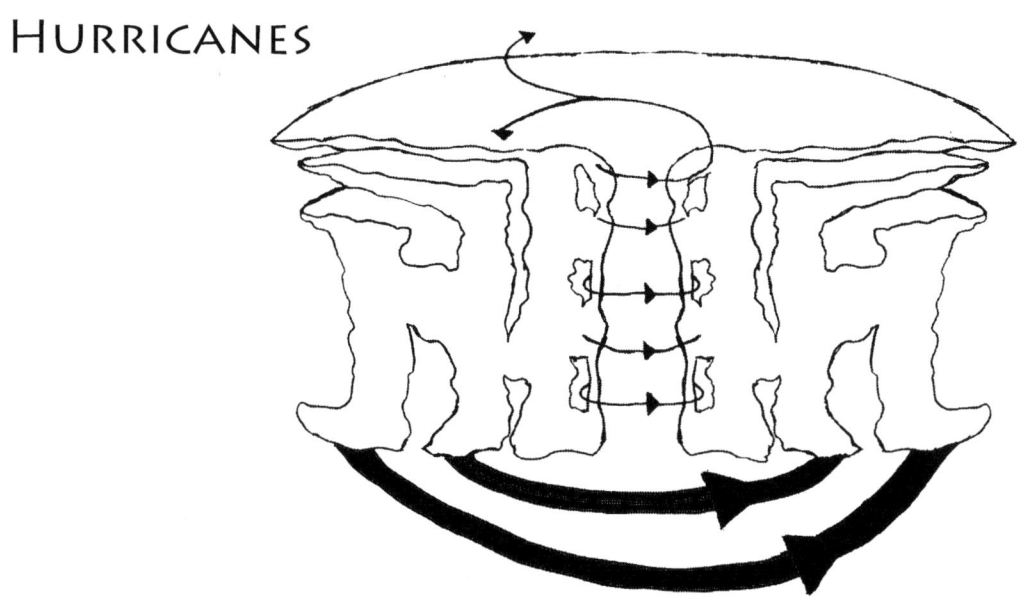

Information Learned

Earth Science Notes
Pacific Ocean

_____
_____
_____
_____
_____
_____
_____
_____
_____
_____
_____

Earth Science Notes
Pacific Ocean

SCIDAT Logbook
Project Record Sheet

Glue Picture of
Project Here

Information Learned

SCIDAT Logbook
Project Record Sheet

Glue Picture of
Project Here

Information Learned

SCIDAT Logbook
Swiss Deciduous Forest Climate Sheet

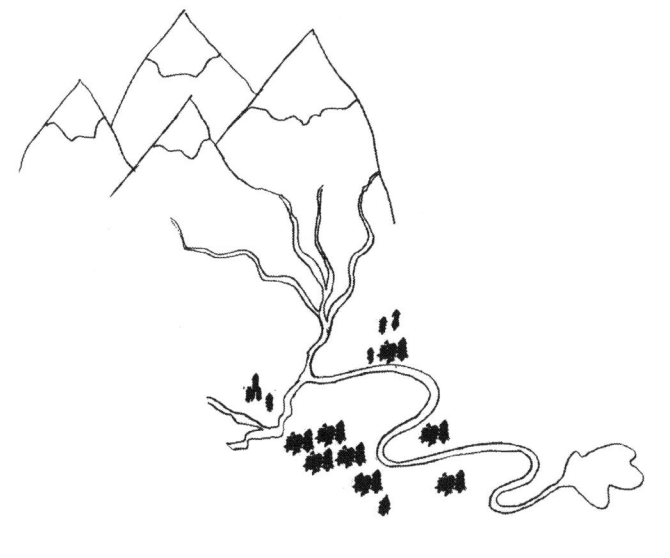

Area Map

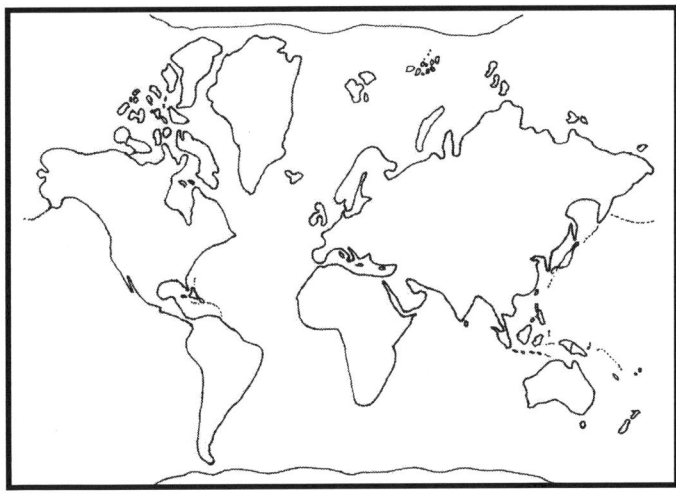

Climate Information

Interesting Facts

Other Types

Weather Records for the Weeks of _____

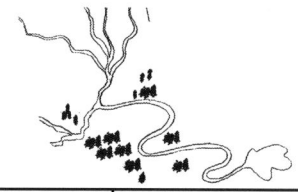

	Monday	Tuesday	Wednesday	Thursday	Friday
High					
Low					
Rainfall					
Conditions					

	Monday	Tuesday	Wednesday	Thursday	Friday
High					
Low					
Rainfall					
Conditions					

SCIDAT Logbook
Earth Science Record Sheet

Groundwater

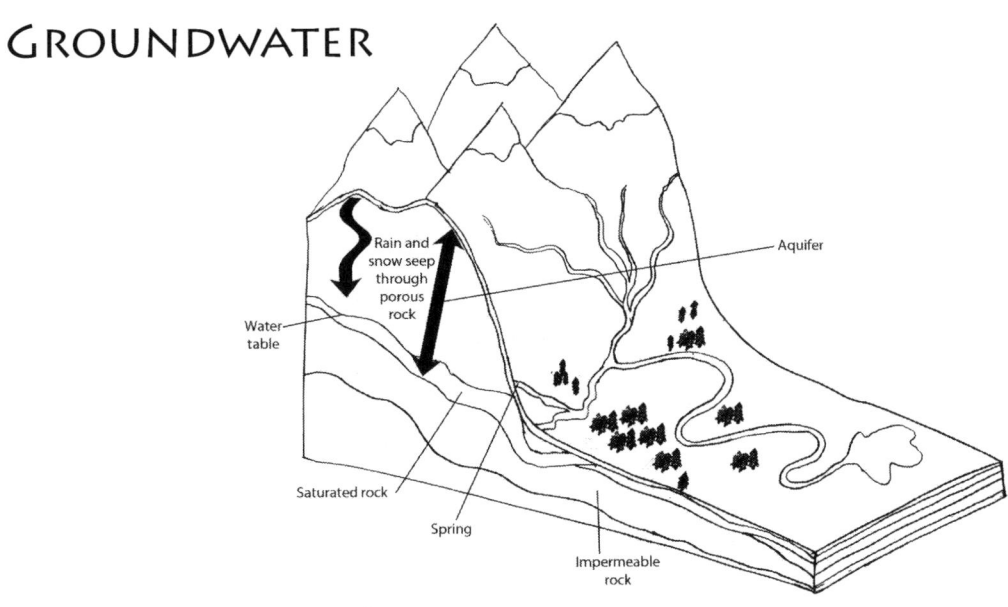

Information Learned

SCIDAT Logbook
Earth Science Record Sheet

Waterfall

Information Learned

SCIDAT Logbook
Earth Science Record Sheet

Rivers

Information Learned

SCIDAT Logbook
Earth Science Record Sheet

Lakes

Information Learned

Earth Science Notes
Swiss Deciduous Forest

Earth Science Notes
Swiss Deciduous Forest

SCIDAT Logbook
Project Record Sheet

Glue Picture of
Project Here

Information Learned

SCIDAT Logbook
Project Record Sheet

Glue Picture of
Project Here

Information Learned

Earth Science Notes
Bonus Data

Bonus Data

As humans, our technological advances can hurt our planet. For instance, our need for paper requires cutting down trees. Our need for food and other goods has changed the earth's landscape through farming and mining. Our need for electricity and mass produced products has created pollution and trash that is damaging to our environment.

To help protect our natural

Earth Science Notes
Bonus Data

BONUS DATA

To help protect our natural resources, we can recycle things like paper and plastic. Throw trash away in a trashcan, not on the ground. Plant things, either in our own backyards or as part of a service project in our community. Switch off lights and other electronics when we are not using them. Finally, we can walk or ride our bikes when we are not traveling very far.

Earth Science Glossary

Earth Science Glossary

*1

WEATHER

*1

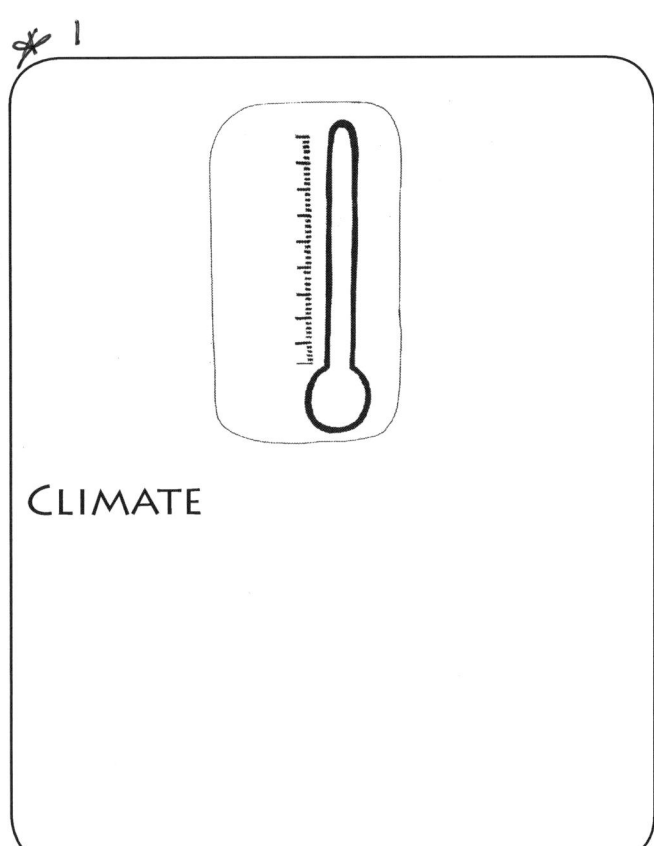

CLIMATE

WIND

GUSTS

Earth Science Glossary

TORNADO

MONSOON

PRECIPITATION

THUNDERSTORM

Earth Science Glossary

Flood

Snowflake

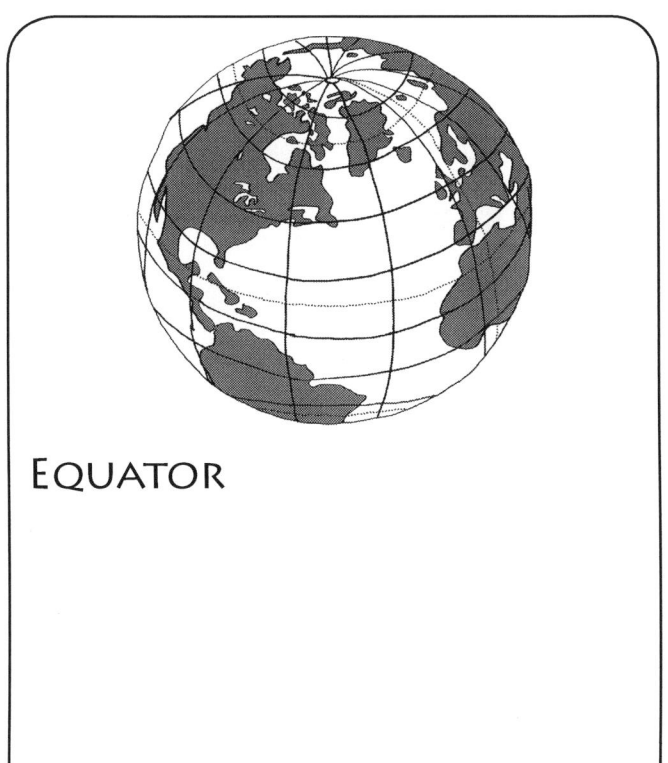

Equator

Season

Earth Science Glossary

Sandstorm

Drought

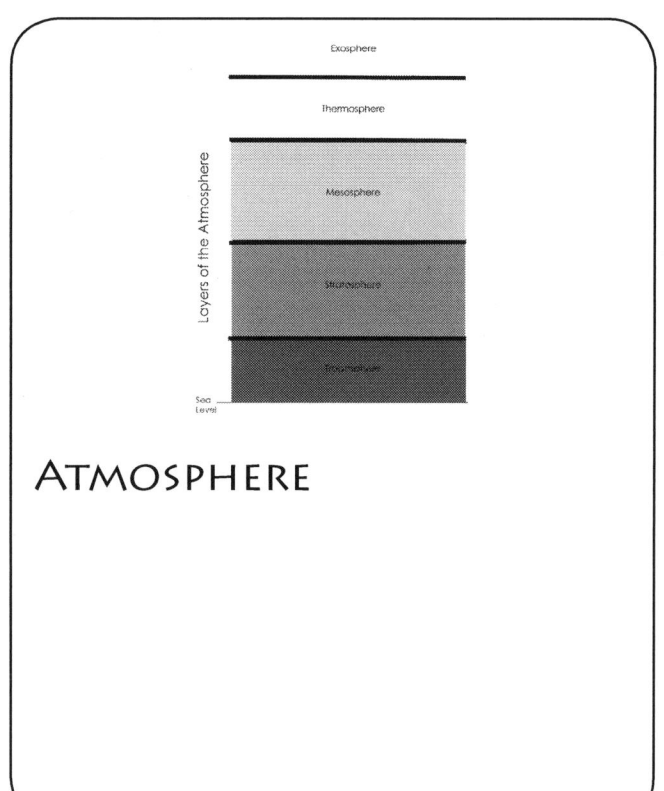

Atmosphere

Vapor

Earth Science Glossary

Clouds

Fog

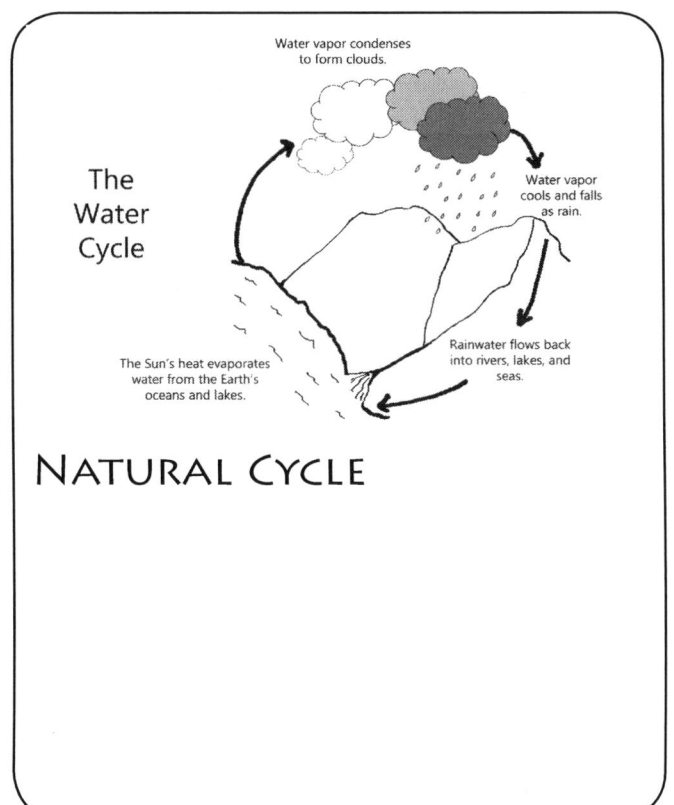

Natural Cycle

Coral

Earth Science Glossary

CURRENTS

AQUIFER

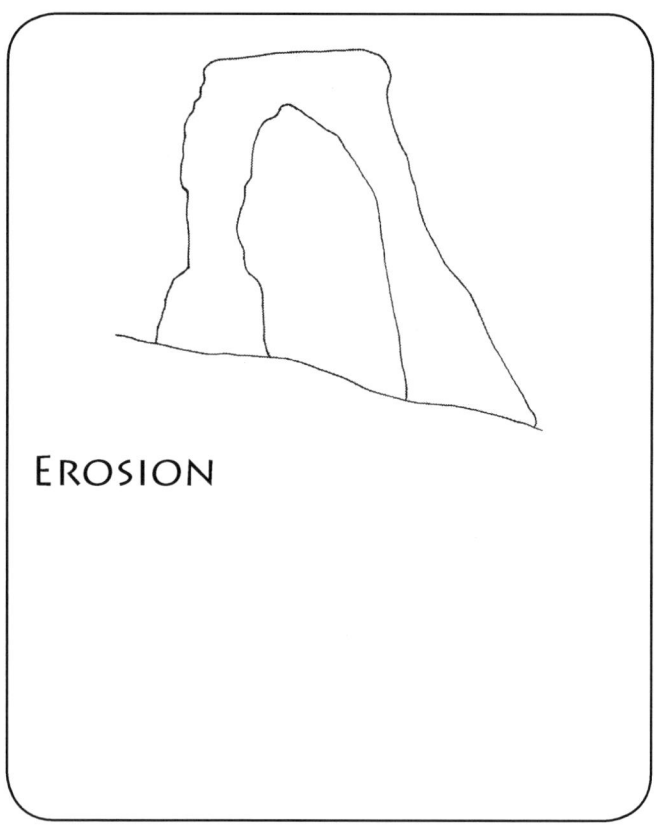

EROSION

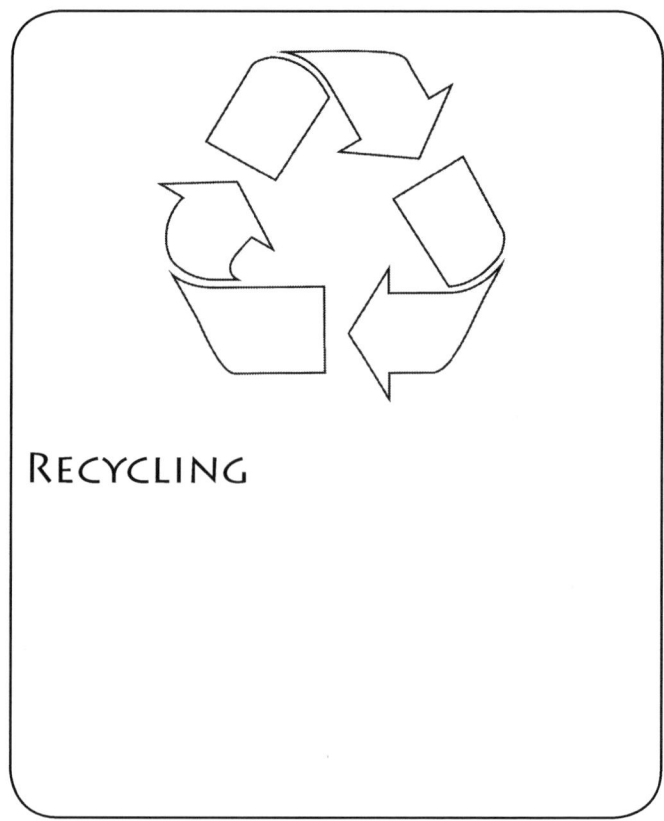

RECYCLING

Made in the USA
Columbia, SC
09 May 2018